"I have many hats," said
Dog. "Pick any one," he said to
Bunny. "You will love these
pretty hats!"

1

Bunny looked at the hats.

"This is a pretty hat," said Bunny. "I'll try it on to see if I like it."

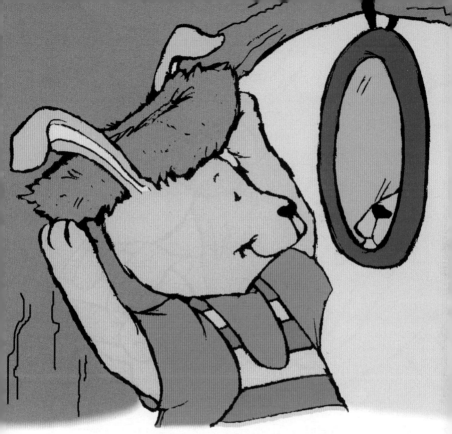

"My, my, my," said Bunny.
"I just **love** this hat. It looks so
pretty on me!"

3

"It is a bit ... let me see ...
lumpy! Try this one," said Dog.

4

"No, no, no," said Bunny.
"This is the hat I want."

She felt a little funny and
dizzy. Things seemed to move.

"My, my," said Bunny. "I
think I need a nap."

Bunny could see two things fly by.

"That's a good spot to put my hat," she said. "I'll just keep it there till next time."

And Bunny went in to
take a nap.

The End